Contents

Each page has a title telling you what it is about.

AT2.5

Letters representing numbers

In the equations on this page each letter represents a number.

$a = 5$ $b = 20$ $c = 45$

$d = 40$ $e = 8$ $f = 10$

Use this information to find the value of the following calculations.

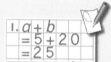

This shows how to set out your work.

1   $a + b =$ 2 $b + c =$

3   $d + c =$ 4 $f - a =$

5   $c \div a =$ 6 $f \times b =$

7   $c \times f =$ 8 $b \times a =$

9   $d \times a =$ 10 $b \div a =$

11   $f \div a =$ 12 $c \div f =$

13   $e + d + f =$ 14 $e + c + e =$

15   $a \times f \times b =$ 16 $e \times f \times f =$

17   Now make up 4 of your own equations for a friend to solve.

Instructions look like this. Always read these carefully before starting.

In your group, use the letters a to f, and the operations +, −, × and ÷ to make as many of the numbers from 0–100 as possible. Record your working.

These are Rocket activities. Ask your teacher if you need to do these questions.

Read this to check you understand what you have been learning on the page.

16   I know that a letter can be used to represent a missing number

Number sequences

Continue each sequence for four more numbers, counting in 25s.

1. 125, 150, 175, 200

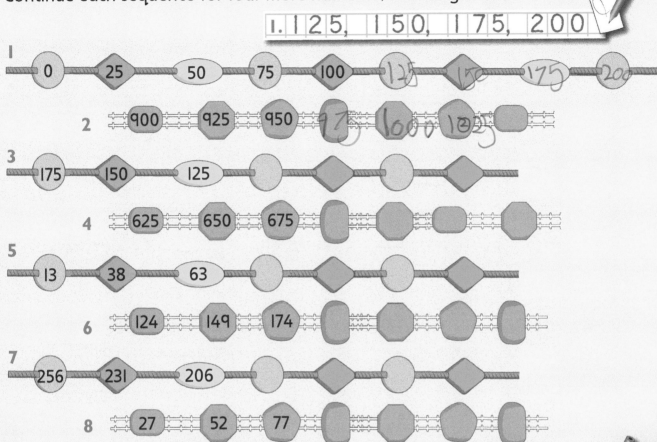

1. 0, 25, 50, 75, 100, 125, 150, 175, 200

2. 900, 925, 950, 975, 1000, 1025

3. 175, 150, 125,

4. 625, 650, 675,

5. 13, 38, 63,

6. 124, 149, 174,

7. 256, 231, 206,

8. 27, 52, 77,

Write the difference between each pair of numbers in the sequences. Write the next three numbers.

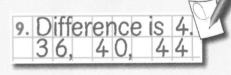

9. Difference is 4.
36, 40, 44

9. 16 20 24 28 32

10. 7 10 13 16

11. 21 26 31

12. 54 60 66

13. 22 27 32

14. 21 15 9

15. 12 9 6

16. 10 7 4

Count up in 8s, starting at 0. How many steps does it take to get past 100? How about past 200?

I can continue a pattern which goes up or down in equal steps

3

Exploring number sequences

Continue each sequence for five more steps.

1
15 40 65 ☐ ☐ ☐ ☐ ☐

2
15 30 45 ☐ ☐ ☐ ☐ ☐

3
11 22 33 ☐ ☐ ☐ ☐ ☐

4
144 132 120 ☐ ☐ ☐ ☐ ☐

5
16 32 48 ☐ ☐ ☐ ☐ ☐

6
8 15 22 ☐ ☐ ☐ ☐ ☐

7
38 32 26 ☐ ☐ ☐ ☐ ☐

8
17 22 27 ☐ ☐ ☐ ☐ ☐

Find the smallest positive number that could have started each sequence.

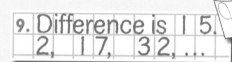

9. Difference is 15.
   2, 17, 32, ...

9   ..., 32, 47, 62, 77, ...

10   ..., 17, 22, 27, ...

11   ..., 22, 28, 34, ...

12   ..., 19, 25, 31, ...

13   ..., 23, 27, 31, ...

14   ..., 38, 43, 48, ...

15   ..., 29, 32, 35, ...

16   ..., 31, 34, 37, ...

In which of the following sequences does the number 272 occur: counting on in 17s from 0 or in 18s from 0? Write two more sequences starting at 0 that contain the number 272.

I can describe the rule of a pattern which goes up or down in equal steps

Fraction sequences

Write the next four numbers in each sequence:

1 2 $2\frac{1}{2}$ 3 $3\frac{1}{2}$ $4.$

2 $\frac{4}{10}$ $\frac{5}{10}$ $\frac{6}{10}$ $\frac{7}{10}$ $\frac{8}{10}$

3 $1\frac{1}{10}$ $1\frac{3}{10}$ $1\frac{5}{10}$ $1\frac{7}{10}$ $1\frac{9}{10}$

4 $1\frac{1}{2}$ $1\frac{3}{4}$ 2 $2\frac{1}{4}$ ...

5 4 $4\frac{1}{2}$ 5 $5\frac{1}{2}$ ...

6 3 $4\frac{1}{2}$ 6 $7\frac{1}{2}$ ...

7 12 $10\frac{1}{2}$ 9 $7\frac{1}{2}$ ...

8 $\frac{1}{3}$ $\frac{2}{3}$ 1 $1\frac{1}{3}$ ...

9 5 $5\frac{3}{4}$ $6\frac{1}{2}$ $7\frac{1}{4}$ ...

10 10 $12\frac{1}{2}$ 15 $17\frac{1}{2}$ ...

Write the missing numbers in these sequences:

11 $2\frac{1}{3}$, $2\frac{2}{3}$, 3,   ____, $3\frac{2}{3}$,   ____

12 $2\frac{1}{4}$, $2\frac{3}{4}$, $3\frac{1}{4}$,   ____,   ____, $4\frac{3}{4}$

13 $3\frac{1}{3}$, $3\frac{2}{3}$, 4,   ____, $4\frac{2}{3}$,   ____

14 $1\frac{1}{2}$,   ____, 3,   ____, $4\frac{1}{2}$,   ____, 6

Invent some fraction sequences of your own, each with one missing number. Invite a friend to find the missing numbers.

I can continue a fraction sequence that goes up or down in equal steps

Number patterns

1. Copy this grid.

				50				

Create a number sequence in which:
- 50 is the middle number
- the difference between next-door numbers is 6
- four numbers come before 50 and four after 50.

2. Repeat this for the following numbers:

Middle number	Difference between next-door numbers
34	7
47	6
5·4	0·5
3·1	0·3
$8\frac{1}{4}$	$\frac{1}{4}$
$5\frac{2}{5}$	$\frac{1}{5}$

Use a hundred square. Colour the multiples of 4 red. Add 2 to each coloured number and colour the answers blue. Describe any patterns you see.

50	51	52	53	54
40	41	42	43	44
30	31	32	33	34
20	21	22	23	24
10	11	12	13	14
0	1	2	3	4	5 6 7 8 9

I can create a number sequence which goes up or down in equal steps

Number sequences

Find the difference between the numbers in each sequence. Write the missing numbers, and the next five numbers in each sequence.

1 100, 89, 78, ☐, 56, ...

2 $4\frac{1}{3}$, ☐, $3\frac{2}{3}$, ...

3 $5\frac{1}{2}$, ☐, ☐, ☐, $6\frac{1}{2}$, ☐, 7, ...

4 3·3, ☐, ☐, 3·6, ...

5 $1\frac{3}{4}$, ☐, $3\frac{1}{4}$, ☐, $4\frac{3}{4}$, ...

6 4·2, ☐, ☐, 7·8, ...

7 $4\frac{2}{3}$, ☐, ☐, $5\frac{2}{3}$, ...

8 3·05, ☐, ☐, 3·08, ...

Invent two missing number sequences, one using fractions, the other using decimals.

These number sequences have been made by adding on 2-digit numbers. The unit digits are the only numbers still showing. What could the step size be in each sequence?

9. Step size 2 5.
16, 41, 66, 91, ...

9 ⬤6, ⬤1, ⬤6, ⬤1, ...

10 ⬤3, ⬤7, ⬤1, ⬤5, ⬤9, ⬤3, ⬤7, ⬤1, ⬤ ...

11 ⬤8, ⬤8, ⬤8, ⬤8, ...

12 ⬤9, ⬤5, ⬤1, ⬤7, ⬤3, ⬤9, ⬤5, ⬤ ...

There is more than one answer for these questions.
Write 2 more possible step sizes for each sequence.

Number patterns

Copy and complete these patterns. Write what the rule is each time.

1   2   5   11   23   47   ☐   ☐   ☐

2   1   2   4   7   11   16   ☐   ☐   ☐

3   1   2   4   8   ☐   ☐   ☐   ☐

4   10   12   16   22   30   40   ☐   ☐   ☐

5   1   4   9   16   ☐   ☐   ☐   ☐

6   75   70   64   57   49   ☐   ☐   ☐

7   20   40   80   160   ☐   ☐   ☐   ☐

8   20   23   29   38   50   ☐   ☐   ☐

9   Create some different number patterns for your partner to describe and continue.

Use a hundred square.
Colour the numbers
1, 13 and 25.
Find the pattern and
complete the sequence.
What do you notice
about the pattern?

30	31	32	33	34	35
20	21	22	23	24	25
10	11	12	13	14	15
0	1	2	3	4	5	6 7

I can continue a pattern with different sized steps

Function machines

Copy and complete the table for each function machine.

1

+350

In	425		382		675	
Out		550		780		1248

2

−425

In	650		518		1386	
Out		375		686		276

These tables show the input and output of one machine.
What was the function each time?

3

In	Out
6	18
4	12
7	21

4

In	Out
189	21
72	8
468	52

5

In	Out
365	240
578	453
1024	899

6

In	Out
475	820
683	1028
345	690

Draw your own function machine but keep its function secret. Tell your partner three inputs and outputs. Can they work out the secret function?

7 Copy and complete the tables.

In	Function	Out
16	halve	8
260	halve	
150	halve	
326	halve	

In	Function	Out
32	quarter	8
	quarter	20
500		125
	quarter	86

I can try different inputs to help me work out the function.

Function machines

Copy and complete the table for each function machine.

1

In	Out
3	
6	
8	
9	
12	

2

In	Out
4	
6	
8	
14	
22	

3

In	Out
3	
12	
7	
5	
24	

4

In	Out
4	
16	
10	
26	
6	

 Use number cards 2, 4, 5, 6 and × to create different inputs and functions. Explore the different outputs the machine produces.

In	Function	Output
3	× 4	8

5

In	40	70	6	80	900
Out					

6

In	5	7	6	8	20
Out					

 I can enter different values into a function machine and work out the outputs

10

Equal (=) or not equal (≠)?

Both ends of this see-saw are equal and balanced.

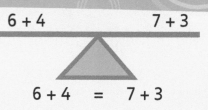

6 + 4 = 7 + 3

But these ends are not equal.

6 + 7   7 + 1

6 + 7 ≠ 7 + 1

Copy and complete these calculations. Put in the correct sign, either = or ≠.

1   23 + 18 ☐ 26 + 17

2   186 – 45 ☐ 174 – 31

3   64 + 19 ☐ 39 + 44

4   351 – 135 ☐ 374 – 158

5   73 + 52 ☐ 94 + 21

6   408 – 75 ☐ 524 – 123

Put numbers in the boxes to make the equations balance.

7   18 + 35 = 26 + ☐

8   724 – ☐ = 686 – 19

9   14 + 52 = 98 – ☐

10   625 – 586 = ☐ – 29

11   363 + ☐ = 258 + 251

12   433 – 394 = ☐ – 128

Find some numbers that could go in the clouds to make the ends balance.

347 + = 286 + ☁☁

521 – = 369 – ☁☁

I can check whether the two sides of an equation balance

11

Balancing equations

Make these equations balance.

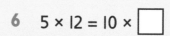

$$3 \times 4 \ = \ 2 \times 6$$

1   $5 \times \boxed{} = 4 \times 10$

2   $10 \times \boxed{} = 4 \times 5$

3   $7 \times 6 = 3 \times \boxed{}$

4   $6 \times 4 = 3 \times \boxed{}$

5   $5 \times 8 = 2 \times \boxed{}$

6   $5 \times 12 = 10 \times \boxed{}$

Use the correct signs to show whether these equations balance.

7   $63 \div 7 \boxed{} 45 \div 5$

8   $72 \div 8 \boxed{} 36 \div 6$

9   $64 \div 8 \boxed{} 42 \div 6$

10   $56 \div 7 \boxed{} 24 \div 3$

11   $240 \div 3 \boxed{} 160 \div 2$

12   $350 \div 7 \boxed{} 400 \div 8$

13   $48 \div 4 \boxed{} 5 \times 3$

14   $120 \div 10 \boxed{} 6 \times 2$

What do you notice about the numbers involved in these equations?

$50 \times 8 = 100 \times 4$ $10 \times 6 = 20 \times 3$

$8 \times 4 = 16 \times 2$ $50 \times 4 = 100 \times 2$

Write three more equations which work in this way.

I can check whether the two sides of an equation balance

Greater than, less than

Copy and complete these equations using > or < .

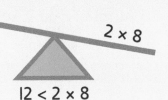

12 < 2 × 8

1   350 ⬜ 30 × 9

2   12 ⬜ 49 ÷ 7

3   40 × 3 ⬜ 140

4   36 ⬜ $\frac{1}{2}$ of 80

5   600 ÷ 3 ⬜ 2

6   67 ⬜ 295 ÷ 5

7   565 ÷ 5 ⬜ 103

8   43 × 3 ⬜ 119

For each box, choose three numbers that could go in it.

9   ⬜ > 63 × 2

10   ⬜ < 420 ÷ 4

11   86 ÷ 2 > ⬜

12   72 × 5 > ⬜

For each box, choose three numbers that could go in it.

13   13 + ⬜ < 19

14   ⬜ < 248 ÷ 4

15   137 + ⬜ < 145

16   200 × ⬜ > 840

17   330 > ⬜ × 55

18   210 > ⬜ × 42

Put numbers in the cloud and the box. The box number must be even and the cloud number must be odd. What possible answers could you have?

326 + ☁ < 337 − ⬜

Number pairs

A number pair is any two numbers that add together to make a total.

Amber and Jack are listing all the number pairs for a total of 4.

1 and 3 is a number pair for a total of 4.
We know 1 + 3 = 4, so we know 3 + 1 = 4.

So 1 + 3 and 3 + 1 are not different number pairs because they use the same numbers.

Amber and Jack find three number pairs for a total of 4:

$$0 + 4 \qquad 1 + 3 \qquad 2 + 2$$

Make a list of all the number pairs for a total of:

1	5	2	6	3	7
4	8	5	9	6	10

How many number pairs are there for each total?

Predict how many number pairs there will be for a total of:

7	11	8	12	9	13	10	14	11	15

12 Why does this pattern of numbers happen?

13 What is the rule for how many number pairs there are for any total?

There are three number pairs for a total of 4.
There are also three number trios for a total of 4:

$$0 + 1 + 3 \qquad 0 + 2 + 2 \qquad 1 + 1 + 2$$

Investigate number trios for other numbers up to 10.

I know that a letter or symbol can be used to represent a missing number

AT2.5

Each shape represents a different number.

100

△
25

■
500

★
75

Use this information to work out the following calculations.

1 + △ =

2 ○ + △ =

3  ★ + ○ + ■ =

4 △ + ★ + ○ =

5 ○ × △ =

6 ★ × ○ =

7 ○ ÷ △ =

8 △ ÷ ○ =

9 ■ + ★ − ○ =

10  ■ + △ + ★ − ○ =

Create your own questions using symbols to represent numbers. Challenge a friend to do the calculations.

Letters representing numbers

In the equations on this page each letter represents a number.

$a = 5$ $b = 20$ $c = 45$

$d = 40$ $e = 8$ $f = 10$

Use this information to find the value of the following calculations.

1. $a + b =$ **25**
   $$= 5 + 20$$
   $$= 25$$

2. $b + c =$ **65**

3. $d + c =$ **85**

4. $f - a =$ **5**

5. $c \div a =$

6. $f \times b =$ **30**

7. $c \times f =$

8. $b \times a =$ **25**

9. $d \times a =$

10. $b \div a =$

11. $f \div a =$

12. $c \div f =$

13. $e + d + f =$

14. $e + c + e =$ **61**

15. $a \times f \times b =$

16. $e \times f \times f =$

17. Now make up 4 of your own equations for a friend to solve.

In your group, use the letters a to f, and the operations +, −, × and ÷ to make as many of the numbers from 0–100 as possible. Record your working.

 I know that a letter can be used to represent a missing number

Solving simple equations

Find the value of *n* in each equation.

1.	$n + 5 = 16$
	$n + 5 - 5 = 16 - 5$
	$n = 11$

1 $n + 5 = 16$

2 $n + 8 = 25$

3 $34 = n + 17$

4 $48 = n + 22$

5 $45 = n + 29$

6 $64 = 48 + n$

7 $75 - n = 50$

8 $38 - n = 22$

9 $65 - n = 39$

10 $18 = n - 7$

11 $26 = n - 9$

12 $58 = n - 17$

Make these balance:

13

$7 + n \quad = \quad 23 - 5$

14

$16 + 9 \quad = \quad 100 \div n$

15

$48 \div 8 \quad = \quad 38 - n$

16

$3 \times 5 \quad = \quad n \div 2$

17

$n \times 10 \quad = \quad 45 + 35$

18

$25 + n \quad = \quad 99 \div 3$

Substitute 0 (zero) for *a*. Then solve each equation in turn. What value do you end up with for *e*?

$a + 1 = b$ $\qquad$ $b + 9 = c$ $\qquad$ $c \div 2 = d$ $\qquad$ $d - a = e$

Repeat this using your own value for *a*. What do you notice?

Create a similar challenge for your partner to solve using letters to represent values.

Growing patterns

Talk about these patterns with a friend. Try to work out a rule to describe what is happening in each pattern. Then write it down.

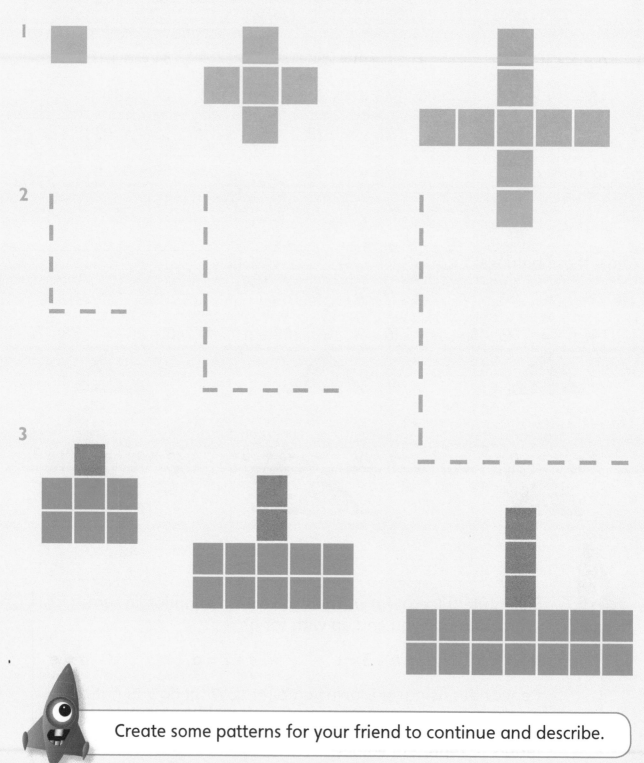

1

2

3

Create some patterns for your friend to continue and describe.

I can work out the rule about how a pattern changes

Square numbers

Draw a picture on squared paper to show how these pegs can be arranged in a square shape.

1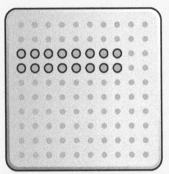

2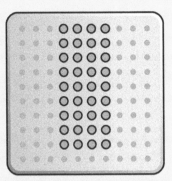

3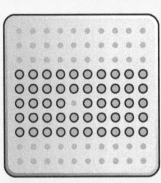

4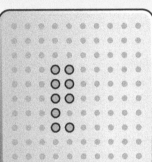

5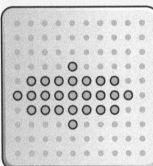

6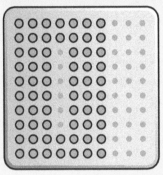

Copy and complete.

7 $4^2 =$

8 $7^2 =$

9 $2^2 =$

10 $9^2 =$

11 $6^2 =$

12 $8^2 =$

13 $5^2 =$

14 $1^2 =$

15 $3^2 =$

16 $10^2 =$

17 $0^2 =$

18 $11^2 =$

Use a set of 1–9 digit cards.

Make different square numbers.

| 3 | 6 | | 4 | 9 | | 1 |

five cards used

How many can you make?
How many of the cards can you use?

I can use the notation of square numbers

Square numbers

Copy and complete.

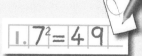

1 $\square^2 = 49$ 2 $\square^2 = 4$ 3 $\square^2 = 100$

4 $\square^2 = 9$ 5 $\square^2 = 64$ 6 $\square^2 = 1$

7 $\square^2 = 81$ 8 $\square^2 = 25$ 9 $\square^2 = 36$

10 Copy and complete this table of squares.

10^2	20^2	30^2	40^2	50^2	60^2	70^2	80^2	90^2	100^2
100									

32 is called a happy number because:

$$32 \longleftarrow \text{Square each digit then add}$$
$$9 + 4 = 13$$
$$1 + 9 = 10$$
$$1 + 0 = 1$$

Repeat until you arrive at a 1-digit number

If you arrive at 1, then the start number is happy.
Investigate how many happy numbers you can find up to 100.

I can remember or work out square numbers

Notation for square and cube numbers

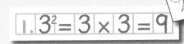
$$1. \quad 3^2 = 3 \times 3 = 9$$

A square number is shown with a small 2.
For example, 3^2 means 3×3 and it has a value of 9.

A cube number is shown with a small 3.
For example, 4^3 means $4 \times 4 \times 4$ and it has a value of 64.

What do the following mean and what is the value of each?

1 3^2	2 5^2	3 4^2	4 6^2
5 3^3	6 5^3	7 4^3	8 6^3
9 2^3	10 1^3	11 10^3	12 9^3

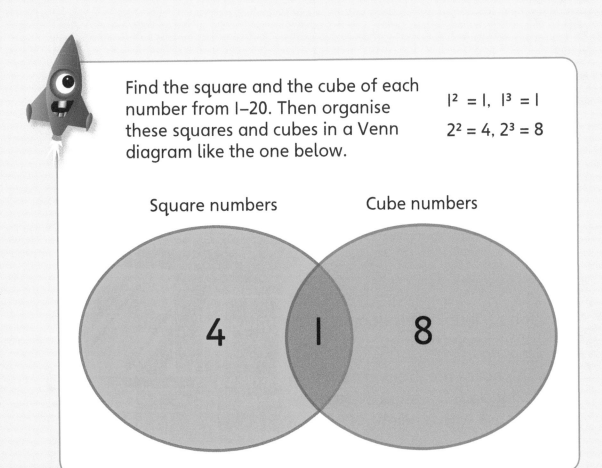

Find the square and the cube of each number from 1–20. Then organise these squares and cubes in a Venn diagram like the one below.

$1^2 = 1, \quad 1^3 = 1$

$2^2 = 4, \quad 2^3 = 8$

Square numbers   Cube numbers

4   1   8

Triangular number patterns

I Copy and complete the following table.

Write some sentences about the patterns you can see.

Spots pattern	Number of dots in each row	Total of dots	How many were added?
●	1	1	
● ● ●	1 2	3	2
● ● ● ● ● ●			
● ● ● ● ● ● ● ● ● ●			

John is helping his dad make a display of boxes in the shop.

The boxes are 10 cm in height. He says to his dad: 'To fit all 21 boxes we need a space with a height of 60 cm.'

Explain this.

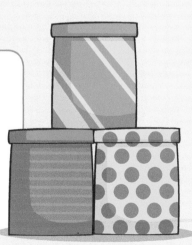

I can recognise and continue triangular number patterns

Your school is hosting a football league and you have to organise the schedule of matches. This is not a knockout tournament. Each team has to play every other team once.

How many matches need to be played if there are:

1 two teams   2 three teams

3 four teams   4 five teams?

5 Look at your results. What pattern can you find?

6 How many games would there need to be for 10 teams? Work this out without writing a list.

Without writing a list, work out how many games there would need to be for:

7 12 teams   8 25 teams   9 42 teams   10 100 teams.

Choose a number of teams for your league. Make a poster containing information about the football tournament.

Whilst you are designing the poster consider the following issues.

- The number of teams who are playing against each other and the order they are playing in.
- How long the tournament will last if each game is 15 minutes.
- Whether teams get a break in-between their matches to rest or if they play again after each match.
- Is your order fair? How can you organise it so teams get a break between matches?
- How the scoring would work.

Triangular number patterns

A new theme park is opening. It is putting all its attractions in a large circle. A path is being built so people can walk directly from one attraction to any other.

If there are two attractions, only one path is needed.

If there are three attractions, three paths are needed.

1   How many paths would be needed if there are four attractions? 4

Recording your workings, work out how many paths are needed for:

2   five attractions 5   3   six attractions 6   4   seven attractions. 7

5   What pattern can you see?

Without drawing a picture, use this pattern to work out how many paths would be needed for:

6   10 attractions 10   7   16 attractions 16   8   55 attractions. 55

Explore how many different journeys someone can make if they walk from each attraction to all of the other attractions. For example, from A to B, B to A, C to A and so on.

What pattern can you see?
How is this different from the paths pattern?

What is the rule for working out the number of journeys for any number of attractions?

How can we use the journeys pattern to help us work out the pathways pattern for any number?

What is the rule for working out the number of paths for any number of attractions?

 I can recognise and continue triangular number patterns

Cube number patterns

 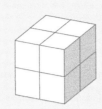 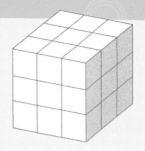

You are building tower blocks of flats. Each flat is a cube shape and each tower block building must be a cube shape, as high as it is wide.

How many flats are there in each tower block when it is:

1 I flat high
2 2 flats high
3 3 flats high
4 4 flats high
5 5 flats high
6 6 flats high?

7 What pattern can you see?

8 What is the rule for how many flats are in a building of any size?

You are going to paint the outside of the tower block buildings. You will paint every side that you can see on the outside of each flat.

 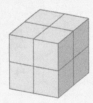

How many squares worth of paint will you use for each tower block when it is:

9   I flat high
10 2 flats high
11 3 flats high
12 4 flats high
13 5 flats high
14 6 flats high?

15 What pattern can you see?

16 What is the rule for how many squares of paint are used for a building of any size?

A tower block costs:
• £200 for every flat in the tower block
• an extra £50 for each square painted on the outside.
Make a poster advertising the prices of four different tower blocks.

I can recognise and continue cube number patterns

25

Pascal's triangle

Work with a partner.

1 Look at the coloured numbers on Pascal's triangle. Can you work out what the numbers all have in common?

2 Take it in turns to say something about other numbers and patterns in the triangle.

3 Calculate the total of the numbers in each row. After a while, try to predict what the next total will be before you calculate it. Are your predictions correct?

3. 1 = 1
   1 + 1 = 2
   1 + 2 + 1 = 4

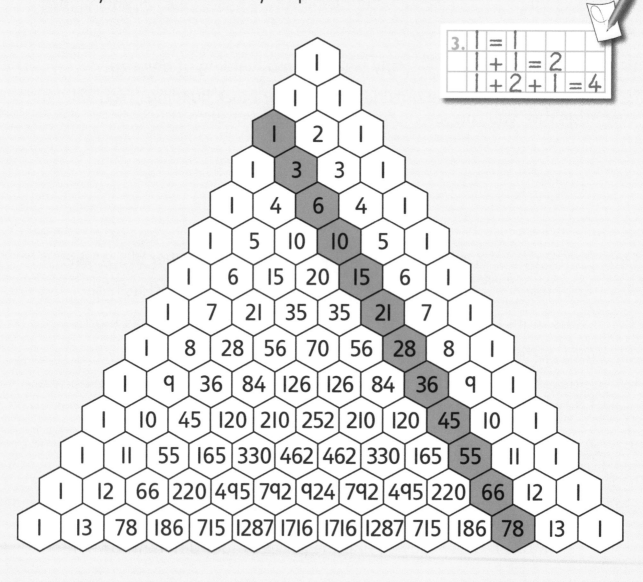

I have investigated the patterns within Pascal's triangle

Fibonacci's pattern

Fibonacci is famous for discovering this pattern of numbers: 1, 1, 2, 3, 5, 8, ...

1   Explore how the numbers in Fibonacci's pattern grow. Can you explain how it works?

2   Continue the pattern up to at least 100.

3   Using the same rule, try pairs of starting numbers other than 1 and 1. Work out the first eight numbers in your pattern. Compare your Fibonacci-type pattern with others in your group.

Complete these Fibonacci-type patterns.

4. 5, 6, 11, 17

4   5, 6, ?, ?

5   2, ?, 12, ?

6   ?, 11, ?, 30

7   1, ?, ?, 21

8   ?, 15, ?, 37, ?

9   7, ?, ?, 39, ?

10 8, ?, ?, ?, 100

Explore the properties of the numbers of the original Fibonacci pattern. Look at where the multiples of 2, 3 and 4 occur. Describe any patterns you find and use these to make a prediction about where the multiples of 5 will occur. Check your prediction.

I can describe the rule for producing a sequence of Fibonacci numbers

Two-step function machines

Patrick was working out the input values for these function machines. But he made some mistakes, and didn't finish his work.

Copy out the tables and check his work. If an answer is wrong or missing put in the correct answer.

1 Function ×3 → +2

In	Out
2	8
5	16
7	1
10	32
	14
	26
	11

2 Function ×4 → −3

In	Out
5	17
3	15
6	21
9	33
	1
	13
	25

3 Function 3n - 4

In	Out
6	14
4	8
8	20
10	26
	29
	11
	17

4 Function 5n + 4

In	Out
3	19
5	21
7	31
10	54
	44
	59
	24

For each machine write down the steps which would reverse the function and return all the output numbers to the original inputs. Check your work by trying out some numbers. For example:

$3n + 2 \rightarrow -2 \div 3$

I can use letters to represent numbers when working with function machines

Two-step function machines

These function machines are set to do two steps.
Copy and complete the tables.

1

In	Step 1: × 10	Step 2: + 7
15	150	157
27		
38		

2

In	Step 1: × 9	Step 2: − 6
20		
35		
76		

3

In	Step 1: ÷ 10	Step 2: + 25
360	36	61
570		
2050		

4

In	Step 1: ÷ 100	Step 2: − 39
48 000		
7500		
6300		

In	Step 1: reverse digits	Step 2 and out: subtract
Machine A 765	567	198

In	Step 1: reverse digits	Step 2 and out: add
Machine B 198	891	1089

Function machine A reverses the input digits then subtracts the new number from the input number.

Function machine B takes this output, reverses the digits and then adds both numbers.

What will these two machines do to 863 and 794?
Try some other numbers. What happens with them?

Two-step function machines

These machines do two functions. First they do Step I, then there are two choices for Step 2. Put a number in the machine and it will come out differently, depending on which Step 2 the machine does.

Note what these function machines do. Then copy and complete the tables.

I

In	Out red	Out blue
8	44	41
14		
32		
50		
11		
17		
22		

2

In	Out red	Out blue
49		
119		
376		
268		
501		
99		
80		

Look at the function machine in question I. Here are some red output numbers. What was the input for each one?

3   56   4   68   5   26   6   596

These are blue output numbers. What number was input each time?

7   53   8   65   9   23   10   227

I can use a function machine that operates on numbers in two steps

Creating expressions

Katie, the waitress, is taking a customer order.

For drinks she could write down *c* and *c* to show that two people want coffee, but it is quicker to write 2*c*. If three people want lemonade she could write 3*l*.

In maths these are known as expressions.

What might Katie have written for:

Write expressions for:

4 apples and 3 pears could be expressed as 4*a* + 3*p*.

Write the following as expressions.

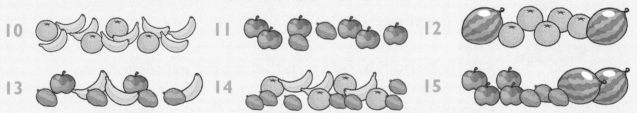

Create your own fruit expressions and ask a partner to work out what they mean.

I can create expressions from pictures

Expressions from number grids

In this grid, 22 is the key number.

21	22	23
$n-1$	n	$n+1$

We can call 22 'n'. Then 23 is worth $n+1$ and 21 is $n-1$.

Copy and complete these grids. Show what each number is worth in relation to the key number.

1

31	32	33
$n+9$	$n+10$	
21	22	23
	n	
11	12	13
$n-11$		

2

63	64	65
$n+9$		
53	54	55
	n	$n+1$
43	44	45

For each of the grids above complete the following tasks.
- Write a chain to add together all the expressions in the grid.
- Simplify your chain to make it as short as possible.
- Now add the numbers in your grid.
- Find a connection between your total number and your expression.
- Does this always work?
- Find some other connections.

I can create expressions from number grids

Triangular and square numbers

1 Complete the table of triangular numbers and how they are made, on PPM 133.

Position	Triangular number	Made by adding
1st	1	1
2nd	3	1 + 2
3rd	6	1 + 2 + 3

2 Complete the table of square numbers and how they are made, on PPM 133.

Position	Square number	Made by multiplying	Made by adding triangular numbers	Positions of triangular numbers
1st	1	1 × 1	1	1st
2nd	4	2 × 2	1 + 2	1st and 2nd
3rd	9	3 × 3	3 + 6	2nd and 3rd

Did you notice that the third square number is the sum of the second and third triangular number, and so on?

3 Can you find a way to show what the nth square number would be?

4 The first two triangular numbers are odd, the second two are even, the third two are odd, and so on. Does this apply beyond the tenth triangular number? If so, why?

Hint: The nth number means any number in the same sequence. To discover a rule for finding any number in the sequence we use n instead of an actual number.

Carl Gauss, at the age of ten, realised how powerful adding consecutive numbers was. What can you find out about him, using books or the internet?

I can identify and predict number patterns

Matchstick patterns

Sequence A

Sequence B

Sequence C

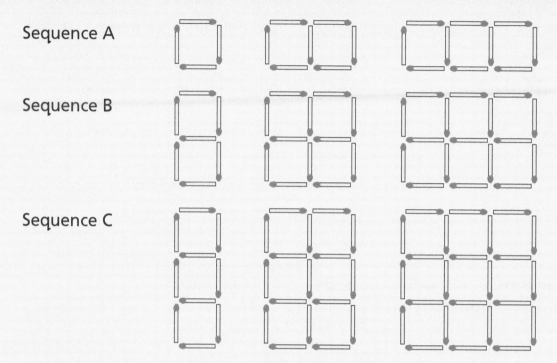

In sequence A you need 4 matchsticks for the first pattern, 7 matchsticks for the second pattern and 10 matchsticks for the third pattern.

1   How many matchsticks would you need for each of the patterns in sequence B? Sequence C?

2   For each sequence A, B and C draw the next two diagrams.
    How many matchsticks would you need for each?

3   Copy and complete this table to show what you have found.

	1st pattern	2nd pattern	3rd pattern	4th pattern	5th pattern
Sequence A	4	7	10		
Sequence B					
Sequence C					

4   For each sequence A, B and C, how many matchsticks would there be in the 10th pattern?

Find a general rule for how each pattern increases.
Then find an expression for the nth shape in each sequence.

I can identify and predict number patterns

Straw patterns

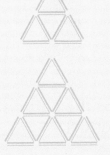

1   How many straws are used to make shape 1?
2   How many triangles have been made?

3   How many straws are used to make shape 2?
4   How many small triangles have been made?

5   How many straws are used to make shape 3?
6   How many small triangles have been made?

7   Complete the table up to shape 10 to show the numbers of straws used and small triangles made. Use straws to make each shape yourself.

Shape	1	2	3	4	5	6	7	8	9	10
Triangular number	1	3	6							
Number of small triangles	1	4	9							
Number of straws	3	9	18							

8   What number patterns can you see in your table?

9   Predict the number of small triangles for shapes 12, 15, 20 and 100.

10 What is the rule for the number of straws and small triangles in shape n? How did you find this rule?

Hint: To find the triangular number, add 1 to the shape number then multiply that by the shape number. Then divide the total by 2.

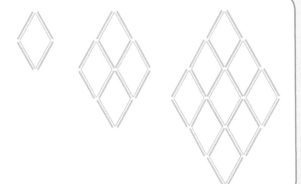

Look at these three patterns.

Draw up a table to find out how many straws you will need for the 10th pattern, the 100th pattern and the nth pattern.

1   Copy this pattern. Continue it for the 4th, 5th and 6th terms.

Position   1   2 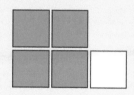   3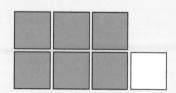

2   Describe the pattern of squares to your partner.

3   Copy and complete this table.

Position	Number of squares
1	3
2	5
3	7
4	
5	
6	

With your partner...

4   Talk about the pattern of results in the table.

5   Agree how the number of squares in a shape is linked to its position.

6   Work out how many squares there would be in the 40th shape. And the 100th shape.

7   If you feed the position number into a function machine, what function must it be for the output to give you the number of squares?

How would the table in question 3 be different if an extra white square was added each time?

What function would the function machine need to do so the output gives you the new number of squares?

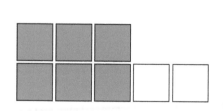

I can describe patterns and identify rules for continuing them

Rules of pattern

1 The coloured dots represent chairs and the rectangles represent dining tables. Show the next two terms in this pattern.

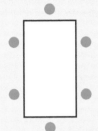 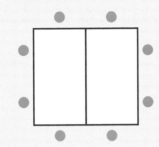 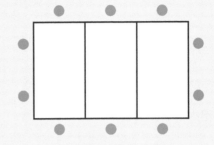

2 Copy and complete this chart.

Number of tables	1	2	3	4	5	6
Number of chairs	6	8	10			

3 Add two columns onto your chart to show the number of chairs for 10 and for 20 tables.

Number of tables	1	2	3	4	5	6	10	20
Number of chairs	6	8	10					

4 Look at the pattern of tables and chairs. No matter how many tables there are always four blue chairs. What is the link between tables and red chairs?

Tables	1	2	3
Red chairs	2	4	6
Blue chairs	4	4	4
Total	6	8	10

 If there were two rows of tables instead of one, how would that affect the numbers of red and blue chairs?

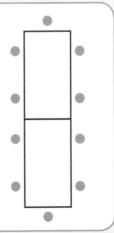

I can identify rules for continuing patterns

Drawing an expression

For the following expressions show a function machine.

1. $n \rightarrow \boxed{\times 3}_{3n} \rightarrow \boxed{+1} \rightarrow 3n+1$

1	$3n + 1$	2	$2n + 3$	3	$4n + 5$
4	$6n - 2$	5	$4n - 3$	6	$6n - 4$
7	$8n - 2$	8	$3n + 7$	9	$4n - 2$

Look at the following inputs and outputs. Which machine in questions 1–9 would each pair have come from?

10	Input 7	Output 25	11	Input 5	Output 18
12	Input 6	Output 25	13	Input 8	Output 37

Choose an input number and operate on it with all these different functions.

Input	Function	Output
	$n + 1$	
	$n + 2$	
	$n + 3$	
	$3n + 6$	
	$5n + 10$	
	$6n + 12$	
	$4n + 12$	

Try putting different numbers through the same functions. Compare your results with a friend's.

8n-2

18

I can solve simple equations by balancing both sides

Inverse functions

Find the inverse of these functions:

1	multiply by 5	2	multiply by 7	3	divide by 4
4	divide by 3	5	add 6	6	subtract 4
7	subtract 325	8	add 735	9	subtract 0·9
10	divide by 20	11	multiply by 26	12	divide by 45

Find the inverse of these functions:

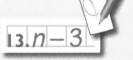

13 $n + 3$ 14 $n + 8$ 15 $n + 6$ 16 $n + 3$

17 $n - 2$ 18 $n - 5$ 19 $n - 4$ 20 $n - 6$

What is the inverse of these functions?

> Hint: Remember $4n$ means $4 \times n$.

21 $5n$ 22 $3n$ 23 $6n$ 24 $18n$

25 $\frac{1}{4}n$ 26 $\frac{1}{8}n$ 27 $n \div 7$ 28 $n \div 10$

Complete these tables. What is the inverse of the functions?

29 Function $2n$

In	Out
4	
8	
12	

30 Function $n + 2$

In	Out
5	
7	
10	

31 Function $\frac{1}{2}n$

In	Out
5	
7	
10	

Write down some functions using n.
Challenge your partner to work out their inverses.

I can work out inverse functions

Solving equations

Draw the function machine
for each equation.
Use inverse operations
working backwards to solve it.

1   $3x + 4 = 16$

2   $2n - 4 = 2$

3   $\frac{1}{2}n + 7 = 13$

4   $\frac{1}{2}s - 5 = 5$

5   $7p + 2 = 142$

6   $3q - 10 = 20$

7   $\frac{1}{4}t + 3 = 8$

8   $\frac{1}{10}m - 15 = 9$

9   $\frac{1}{5}x - 7 = 10$

10   $\frac{1}{4}p + 9 = 13$

If you were teaching someone who did not know how
to solve an equation what would you tell them?
Use a diagram to help you explain.
Compare your way of teaching to your partner's.
Whose method did you prefer?
Do you both prefer the same method?

I can solve simple equations by drawing function machines

Draw the function machine for each equation. Use inverse operations working backwards to solve it.

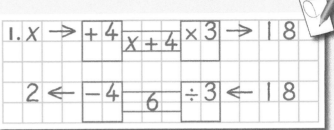

Step 1 +4   Step 2 ×3

1   $3(x + 4) = 18$

2   $2(y + 5) = 18$

3   $3(s + 2) = 36$

4   $\frac{1}{2}(m + 5) = 5$

5   $\frac{1}{2}(n - 10) = 10$

6   $4(b - \frac{1}{2}) = 10$

7   $3(d + 2\cdot5) = 12$

8   $\frac{1}{10}(f - 3) = 1$

9   $5(p - 4\cdot8) = 17$

10   $\frac{1}{4}(q + 16) = 20$

Equations can have unknown numbers on both sides of the equals sign, for example: $4n = 12 + n$.

Draw a diagram like this to help you solve these equations:

$3n = n + 6$ $35 - 2n = 5n$

$4n = 6 + n$ $3n = 16 + n$

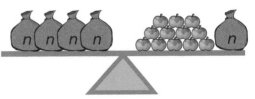

Reversing functions

Stuart and Lorna are talking about their maths.

Lorna says: 'If a function is $2n - 1$ the inverse function is $\frac{1}{2}n + 1$.'

Stuart says: 'That cannot be right.'

1 Copy and complete these tables.

Function $2n - 1$

In n	Step 1: $2n$	Step 2 and out: $- 1$
3	6	5
9	18	17
15		
11		

Function $\frac{1}{2}n + 1$

In n	Step 1: $\frac{1}{2}n$	Step 2 and out: $+ 1$
5		
17		

2 Who is correct? How do the tables help you know this?

Invent some functions using 'n'. Then find the function that reverses each of them.

I can work out how to reverse the changes made by a function machine

Balancing equations

How many cubes are in the cup?
Take 2 cubes away from each side. The equation still balances.
Instead of pictures, we can write equations:

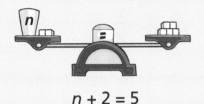

$n + 2 = 5$

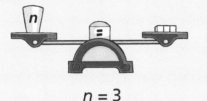

$n = 3$

$$n + 2 = 5$$
$$n + 2 - 2 = 5 - 2$$
$$n = 3$$

Work out the value of n each time.

1

$n + 5 = 13$

2

$n + 8 = 35$

3

$n + 19 = 32$

4

$n + 24 = 53$

5

$n + 46 = 100$

6

$n + 28 = 125$

7

$73 + n = 205$

8

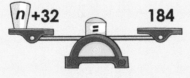

$32 + n = 184$

9

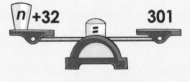

$n + 32 = 301$

The scales have broken so the shop-keeper uses an old-fashioned balance to weigh the fruit. A bag of apples and a 50 g weight balance a 500 g weight.

What is the weight of the bag of apples?

I can solve simple equations by balancing both sides

Balancing equations

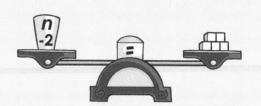

The number of cubes in the cup is *n*, but the cup doesn't balance the 5 cubes.

Remove two cubes from the cup and now the scales balance: $n - 2 = 5$.

How many cubes were in the cup to begin with?
Keep the balance by adding 2 to each side.

$$n - 2 = 5$$
$$n - 2 + 2 = 5 + 2$$
$$n = 7$$

The cup held 7 cubes to begin with.

Find the value of *n* in these equations.
If you want to, sketch a balance to help you.

1 $n - 5 = 4$	2 $n - 8 = 20$	3 $n - 12 = 32$
4 $n - 18 = 47$	5 $n - 19 = 92$	6 $n - 28 = 114$
7 $n - 17 = 245$	8 $n - 58 = 206$	9 $n - 37 = 298$

Simon says: 'I'm thinking of a number. If I take away 17, I'm left with 24.'
What is Simon's number?
Make up some problems like this for a friend to tackle.

I can solve simple equations by balancing both sides

Balancing equations

How many cubes are in the cup?
We can write equations to find the value of n:

$2n = 10$

$$2n = 10$$
$$2n \div 2 = 10 \div 2$$
$$n = 5$$

Work out the value of n each time.

1

18

$2n = 18$

2

24

$3n = 24$

3

30

$2n = 30$

4

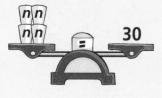

30

$4n = 30$

5

33

$3n = 33$

6

16

$4n = 16$

7

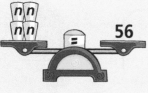

56

$4n = 56$

8

36

$3n = 36$

9

26

$2n = 26$

Draw a diagram to help you answer this question.
A balance has 2 toy lorries and 3 toy cars on one side.
This is balanced by 13 cars on the other side.
How many cars weigh the same as one lorry?

Interpreting problems

Solve these problems in your own way, by showing them as equations. Explain your thinking to a partner.

$$1. \quad 2n + 3 = 17$$
$$2n + 3 - 3 = 17 - 3$$
$$2n = 14$$
$$n = 7$$

1 I think of a number, multiply it by 2 and then add 3 and I have 17. What was my number?

2 I think of a number, multiply it by 3 and then subtract 4 and I have 23. What was my number?

3 I think of a number, multiply it by 5 and then add 6 and I have 26. What was my number?

4 I think of a number, multiply it by 6 and then subtract 8 and I have 52. What was my number?

5 I think of a number, multiply it by 8 and then subtract 9 and I have 79. What was my number?

I think of a number, multiply it by 3 and add on double my number then subtract 5 and I have 15. What was my number?

6 Create your own 'think of a number' problem for your partner.

I can represent and solve word problems by writing equations

Interpreting problems

The table shows part of the football league table at the end of January.
Not all teams have played the same number of games.
These points are awarded to the teams:

- plus 4 points for every match won

- plus 2 points for every match drawn

- minus I point for every match lost.

Team	Won	Drawn	Lost	Points
Lagan	9	0	a	39
Bann	7	4	I	b
Strule	5	c	4	32
Foyle	4	d	9	17
Blackwater	e	6	0	28
Donard	5	f	3	31
Cavehill	g	5	2	48
Mourne	6	h	6	36

I For each team create an equation to calculate the value of the letter. Solve using balancing or by a function machine.

Lagan:
$$(9 \times 4) + a = 39$$
$$36 + a = 39$$
$$36 - 36 + a = 39 - 36$$
$$a = 3$$

Add another two teams, each with a missing score.
Ask your partner to find the value of the missing score.

I can represent and solve word problems by writing equations

Author Team: Peter Gorrie, Lynda Keith, Lynne McClure and Amy Sinclair
Consultant: Siobhán O'Doherty

Published by Pearson Education Limited, a company incorporated in England and Wales, having its registered office at Edinburgh Gate, Harlow, Essex, CM20 2JE. Registered company number: 872828

www.pearsonschools.co.uk

Heinemann is a registered trademark of Pearson Education Limited

Text © Pearson Education Limited 2012

First published 2012

17
10 9 8 7 6 5 4 3 2

British Library Cataloguing in Publication Data
A catalogue record for this book is available from the British Library

ISBN 978 0 435 07766 2

Typeset by Debbie Oatley @ room9design and revised by Mike Brain Graphic Design Limited, Oxford
Illustrations © Harcourt Education Limited 2006–2007, Pearson Education Limited 2010
Illustrated by Fred Blunt, Matt Buckley, Jonathan Edwards, Stephen Elford, Andy Hammond, John Haslam, Sim Marriott, Debbie Oatley, Andrew Painter, Tom Percival, Anthony Rule and Q2A
Cover design by Pearson Education Limited
Cover illustration by Volker Beisler © Pearson Education Limited
Printed in the UK by Ashford Colour Press Ltd.

Acknowledgements
The Publishers would like to thank the following for their help and advice:
Liam Monaghan
Hilary Keane
Stephen Walls

Every effort has been made to contact copyright holders of material reproduced in this book. Any omissions will be rectified in subsequent printings if notice is given to the publishers.